YOUR KNOWLEDGE HAS VALUE

Nuclear chemistry. Methods for the detection of Isotopes and applications of radioactive isotopes

Purvesh Shah

Bibliographic information published by the German National Library:

The German National Library lists this publication in the National Bibliography; detailed bibliographic data are available on the Internet at http://dnb.dnb.de.

ISBN: 9783346526564
This book is also available as an ebook.

© GRIN Publishing GmbH
Nymphenburger Straße 86
80636 München

Print and binding: Books on Demand GmbH, Norderstedt, Germany
Printed on acid-free paper from responsible sources.

The present work has been carefully prepared. Nevertheless, authors and publishers do not incur liability for the correctness of information, notes, links and advice as well as any printing errors.

GRIN web shop: https://www.grin.com/document/1127293

Nuclear Chemistry

Dr. Purvesh Shah

Assistant Professor,

Department of Chemistry,

K.K.Shah Jarodwala Maninanagar Science College,

Ahmedabad-380008

Nuclear chemistry

<u>Content</u>

❖ Introduction	**3**
❖ Detection of isotopes	**10**
❖ Bainbridge velocity focusing mass spectrograph	**10**
❖ Nier's double focusing mass spectrometer	**14**
❖ Applications of Radioactive Isotopes	**16**
❖ Examples	**21**

Nuclear chemistry

Introduction

☐ A nuclear reaction is different from a chemical reaction.

☐ In a chemical reaction, atoms of the reactants combine by a rearrangement of extra nuclear electrons but the nuclei of the atoms remain unchanged.

☐ In a nuclear reaction the nucleus of the atom is involved.

☐ The number of protons or neutrons in the nucleus changes to form a new element.

☐ A nuclear reaction involves a change in the composition of the nucleus.

☐ The number of protons and neutrons in the nucleus is altered.

☐ The product is a new nucleus of another atom with a different atomic number and/or mass number. Thus, a nuclear reaction is one which proceeds with a change in the composition of the nucleus so as to produce an atom of a new element.

Nuclear chemistry

☐ The conversion of one element to another by a nuclear change is called transmutation.

☐ We have already considered the nuclear reactions of radioactive nuclei, producing new isotopes.

❖ **Nuclear Chemistry:** A study of the nuclear changes in atoms is termed Nuclear Chemistry.

❖ **Nuclide:** a term used to refer to a particular atom or nucleus with a specific neutron number N and atomic (proton) number Z.

❖ Nuclides are either stable (i.e., unchanging in time unless perturbed) or radioactive (i.e., they spontaneously change to another nuclide with a different Z and/or N by emitting one or more particles). Such radioactive nuclides are termed radio nuclides.

Nuclear chemistry

The symbol used to denote a particular isotope is

$$^{A}_{Z}X$$

Where X is the chemical symbol and A = Z + N, which is called the mass number.

❖ **Isotopes:**

The atoms of element having same atomic number but different mass number are called isotopes. All isotopes have the same chemical properties.

The isotopes of some elements are the following.

$_1H^1$, $_1H^2$, $_1H^3$

$_2He^3$, $_2He^4$

$_8O^{16}$, $_8O^{17}$, $_8O^{18}$

$_{17}Cl^{35}$, $_{17}Cl^{37}$

$_{92}U^{235}$, $_{92}U^{238}$

Nuclear chemistry

Because isotopes of the same element have the same number and arrangement of electrons around the nucleus, the chemical properties of such isotopes are nearly identical.

Only for the lightest isotopes (e.g., ^{1}H, deuterium ^{2}H, and tritium ^{3}H) are small differences noted.

For example, light water ^1H_2O freezes at 0 °C while heavy water ^2H_2O (or D_2O since deuterium is often given the chemical symbol D) freezes at 3.82 °C.

❖ **Isobar:**

The nuclei which have the same mass number (A) but different atomic number (Z) are called isobars.

Isobars occupy different positions in periodic table so all isobars have different chemical properties. Some of the examples of isobars are $_1H^3$ and $_2He^3$,

$_6C^{14}$ and $_7N^{14}$,

$_8O^{17}$ and $_9F^{17}$.

❖ Isotones:

The nuclei having equal number of neutrons are called isotones.

For them both the atomic number (Z) and mass number (A) are different, but the value of (A - Z) is same.

Examples:

$_4Be^9$ and $_5Be^{10}$,

$_6C^{13}$ and $_7N^{14}$,

$_8O^{18}$ and $_9F^{19}$,

$_3Li^7$ and $_4Be^8$,

$_1H^3$ and $_2He^4$

❖ Isomers:

Atoms of elements which have same atomic number and atomic mass but different radioactive properties(life time) are called isomer.

Nuclear chemistry

e. g. $_{35}Br^{80}$ ($t_{1/2}$ =18 minutes) and $_{35}Br^{80}$ ($t_{1/2}$ =4.5 hours)

The same nuclide (same Z and A) in which the nucleus is in different long lived excited states. For example, an isomer of ^{99}Te is ^{99m}Te where the m denotes the longest-lived excited state (i.e., a state in which the nucleons in the nucleus are not in the lowest energy state).

❖ **Isoelectronic species :**

Atoms and ions that have the same electron configuration are said to be isoelectronic. Examples of isoelectronic species are N^3, O^2, F^- ,Ne,Na^+, Mg^{2+} and Al^{3+} ($1s^2 2s^2 2p^6$).

Another isoelectronic series is P^3, S^2, Cl^-, Ar, K^+, Ca^{2+}, and Sc^{3+} ([Ne]$3s^2 3p^6$).

❖ **Isodiaphers:** In nuclear physics and radioactivity, isodiaphers refers to nuclides which have different atomic numbers and mass numbers but the same neutron excess, which is the difference between numbers of neutrons and protons in the nucleus.

For example,

Element	$^{238}_{92}U$	$^{234}_{90}U$
No. of proton(Z)	92	90
No. of Neutron (A-Z)	238-92=146	234-30=144
Neutron Excess (Neutron-Proton)	146-92=**54**	144-90=**54**

The difference between the neutron number (N) and proton number (Z) is same.

❖ **Mirror nuclei:** Nuclei having the same mass number(A) but with the proton number (Z) and neutron number (A-Z) interchanged (or whose atomic numbers differ by 1 are called mirror nuclei for example. $_1H^3$ and $_2He^3$, $_3Li^7$ and $_4Be^7$.

Nuclear chemistry

❖ **Detection of Isotopes:**

There are various methods for detection of Isotopes:

(1) Thomson's parabola method

(2) Aston's mass spectrograph

(3) Bain-bridge velocity focusing mass spectrograph

(4) Dempster direction focusing (centric) mass spectrograph

(5) **Nier's double focusing mass spectrograph.**

❖ **Bainbridge velocity focusing mass spectrograph**

Bain-bridge velocity focusing mass spectrograph is use to detect isotopes.

A velocity selector was used to produce a monovelocity ion beam and a (transverse) magnetic field was employed to distinguish between ions of different masses.

Nuclear chemistry

- **Construction**

(i) **Ionization Chamber:** Ionization chamber is used to ionize the gas whose mass or isotope is to be determined and positive ions are produced.

(ii) **Velocity Selector:** Velocity selector has two fields electric and magnetic field both are applied perpendicular to the moving ion beam.

(iii) **Vacuum /Analyzing Chamber:** Vacuum/ Analyzing Chamber is a semi-spherical cavity in which another magnetic field is applied perpendicular to the monovelocity positive ion.

Nuclear chemistry

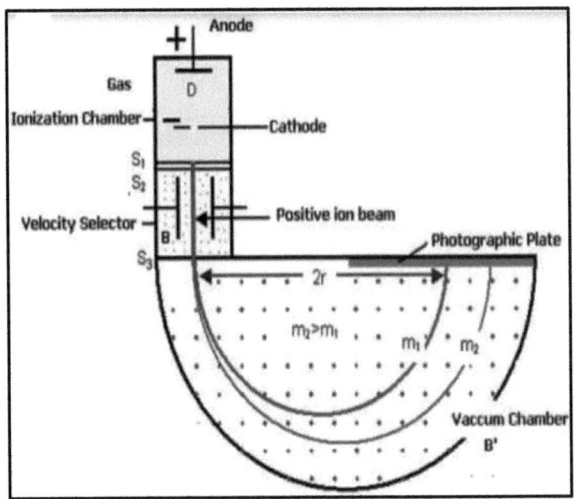

By using electric lamp, the cations produce, which are passed through slits (S_1 and S_2), and then passed through plates (P_1 and P_2).

After they enter into Velocity selector,in which electric field (force) and/or magnetic field (force) are applied in such a way that their effect remain in opposite direction to each other. Under the effect of both these fields only constant velocity containing rays can pass through slit S3 and in magnetic chamber but in magnetic chamber ,there is an existance of magnetic field (force) H, under these effect they occupy circular semicircular path, and strike on photographic plate 'F'. For charged particle following equation is used.

$$\frac{e}{m} = \frac{v}{Hr}$$

Where,

e = charge of particle

m= mass of particle

v = velocity of particle

H = frequency of (magnitude) of magnetic field

r = radius of semi-circular way (path)

In equation (1) e, v and H are constant.

$$\frac{1}{m} = \frac{1}{r}$$

$$\therefore \mathbf{m} \propto \mathbf{r}$$

Thus, elements having higher atomic mass, strike at far end (at m_2) of photographic plate;

while elements having lower atomic mass strike at near end (at m_1) of photographic plate.

❖ Nier's double focusing mass spectrometer:

Nier's double focusing mass spectrometer This spectrometer use to detect isotopes.

It is double focusing spectrometer, which is focuse direction and velocity.

In this method, there are three main parts present as follow:

1.   Ion source
2.   Copper tube and electromagnet
3.   Ion collector

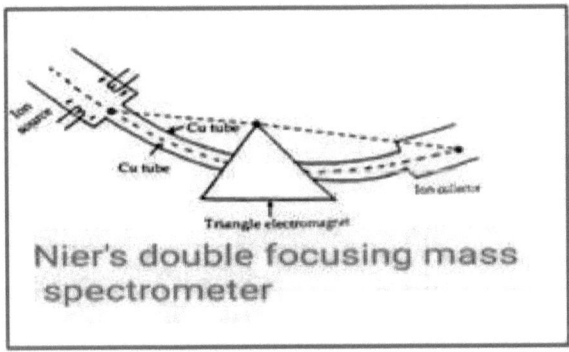

Nier's double focusing mass spectrometer

Nuclear chemistry

As shown in figure cation is produce from electric lamp and they passing through slit and come into in copper (Cu) tube, which is attached with electromagnet. The cation after passing through copper tube, they focused in ion-electro Storage,where they produce current, which is measured by electrometer. Thus the value of potential is changed and raising current is measured. (noted). Then plot the graph of ion current vs. applied potentials.

This plot shows different peaks. These peaks, gives related amount of isotopes along with electricity. In the following plot total five isotopes are present which have different percentage amount.

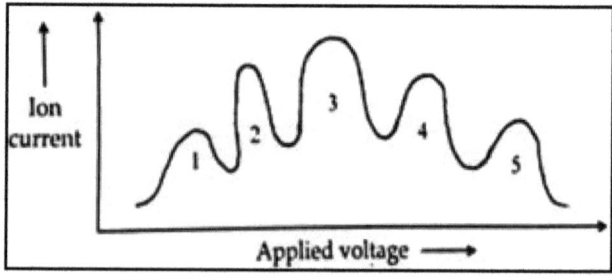

Nuclear chemistry

❖ **Applications of Radioactive Isotopes**

Tracer technique: Application of radioactive isotope is in use through Tracer technique.

(1) To find out solubility of semi-soluble (or partly) salts:

To find out solubility of partly soluble salt radioactive isotopes are use. For example, here we take partly soluble salt $PbSO_4$. Exact amount of radioactive Lead (Pb^*) is mixed with known soluble amount of Lead (Pb). The solution of this mixture is made in dil. HNO_3. Add H_2SO_4 to above solution, we obtained precipitates of $PbSO_4$ is obtained.

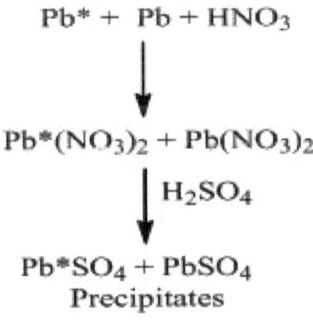

$$Pb^* + Pb + HNO_3$$
$$\downarrow$$
$$Pb^*(NO_3)_2 + Pb(NO_3)_2$$
$$\downarrow H_2SO_4$$
$$Pb^*SO_4 + PbSO_4$$
Precipitates

Precipitates of Lead sulphate are filtered and measured radioactivity by using Geiger Muller radioactivity instrument. During radioactivity amount of radioactive lead in solution is obtained.

If the amount of both the kinds of lead taken initially is known then the amount of non-radioactive lead is measured.

(2) In the study of different chemical reaction:

(a) Photosynthesis: In photosynthesis, leaves of green plant, in presence of sunlight prepare glucose and oxygen by using carbon dioxide and water. The oxygen of glucose is obtained from CO_2, which is confirmed by Tracer technique.

(b) In Friedle-Craft reaction: In Friedle Craft reaction, benzene and acetyl chloride react in presence of anhydrous $AlCl_3$ and gives Acetophenone.

$$C_6H_6 + CH_3COCl + \text{Anhy. } AlCl_3{}^* \longrightarrow C_6H_5COCH_3 + AlCl_3 + HCl^*$$

In this reaction $[CH_3CO][AlCl_4{}^*]$ compound is formed and chlorine is separated from it.

(3) In the field of Medical Science:

The exact position of affective part in body can be identified by radioactive iostopes. in case of blood circulation is not proper occur in body, then radio isotope of sodium (^{24}Na) containing injection of NaCl is given and at exact position where dissolution of NaCl resisted, is known.

Similarly, to decide exact position of tumour during the operation of brain tumour, I^{131} (radioactive Iodine) is used. for treatment of some diseases like cancer radio isotopes are also used. i.e. For the treatment of cancer γ-radiation obtained from Co^{60} is used.

(4)   To find out age of archaeological as well as geological specimen:

The carbon dioxide contain constant amount of C^{14}. During photosynthesis green plant used CO_2 . So specific amount of C^{14} transfer into green plants, but according to natural arrangement whatever the amount of C^{14} is added in green plant during photosynthesis, the similar amount leave by decomposition. When green plant is cut off or converted into dead organic element such as cotton, wood ,etc. new C^{14} is not added but decomposition of C^{14} remain nonstop so the amount of C^{14} is reduced. By evaluating the radioactivity of live wood and dead wood, with known half life time of C^{14} (5600 years) reducing radioactivity of specimen can be calculated. The almost 5000 years old specimen is found through this method.

(5) In Agriculture:

By using radioactive isotopes in agriculture field, we used to decide the application of different fertilizers for different agricultural task, and at which time, what kinds of fertilizers are used.

Take the example of phosphorous containing fertilizers. By applying this method it is prove that during plantation of maize (farming) using phosphorous containing fertilizers can be more beneficial, After growth of crop, this fertilizer is not helpful, while in potatoes continuously supply of phosphorous is required, but for tobacco phosphorous is not used. This can be confirmed through radioactive isotope of phosphorous (P*).

In addition, this method can be applicable in the study of absorption, in qualitative analysis in industries, in identification of metal, in the study of surface area of crystal, radio isotopes.

Nuclear chemistry

Ex.1 For modern mass spectrograph r=15cm and H=0.2 gauss/cm are general values. Calculate the accelerated voltage to identify ^{20}Ne ions.

(M_{Ne} = 3.332 x 10^{-24} gm).

solution : Accelerated voltage for modern mass spectrograph,

$V=(r^2H^2/2m)e$

where, r =15 cm, H = 0.2 gauss/cm,

Mass M_{Ne} = 3.32 x 10^{-24} gm

e = electron charge 1.602 x 10^{-19} coulomb

$$V=\frac{\{(15)^2 \times (0.2)^2 \times 1.602 \times 10^{-19}\}}{\{2 \times 3.32 \times 10^{-24}\}}$$

V= 2.17 x 10^5 volt/cm

Nuclear chemistry

Ex.2. **In mass spectrograph of Dempster, 3000 gauss magnetified and 50 cm path radius is used. Focusing (i) H^+ ions, (ii) Na^+ ions in ion storage. How much accelerated voltage is required ? ($M_{Na} = 23$)**

Solution : Accelerated voltage for modern Dempster mass spectrograph,

$V = (r^2 H^2/2m)e$

where, V = accelerated voltage

       H = magnetic field = 3000 gauss

       r = radius = 50 cm

       m = molecular mass

(i) Accelerated voltage for H^+ ion:

$V = \{(50)^2 \times (3000)^2 \times 4.825 \times 10^{-5}\}/\{1\}$ ---------

$[M_H{}^+ = 1, 1ev = 96.4905KJ$ mole, $1ev = 9.64905 \times 10^{-5}$ and $e/2 = 4.825 \times 10^{-5}$

V = 1.0858 volt

(b)   Accelerated voltage for Na^+ ions

$V = \{(50)^2 \times (3000)^2 \times 4.825 \times 10^{-5}\}/\{23\}$ ---------

$[M_{Na}{}^+ = 23, 1ev = 96.4905KJ$ mole,

$1ev = 9.64905 \times 10^{-5}$ and $e/2 = 4.825 \times 10^{-5}]$

Nuclear chemistry

V= 471.74 volt

Ex. 3. Calculate heat energy released in the fallowing nuclear reaction:

$^2_1H = ^3_1H + ^1_1H$

[Mass of ^1_1H = 1.00782, ^2_1H = 2.014092, ^3_1H = 3.0160441]

Solution:

(a) Reduction of mass = Mass of reactants - Mass of product

$= 2$ (Mass of ^2_1H) - (Mass of $^3_1H + ^1_1H$)

$= 2 \times 2.014092 - (3.016044 + 1.00782)$

$= 4.028184 - 4.023864$

$\Delta m = 0.00432$

(b)   Now released energy,

$\mathbf{\Delta E = \Delta m \, c^2}$

$= 0.00432 \times (3 \times 10^{10})^2$

$= 4.32 \times 10^{-3} \times 9 \times 10^{20}$

$= 4.32 \times 9 \times 10^{17}$

$= 38.88 \times 10^{17}$

$\mathbf{\Delta E = 3.888 \times 10^{18} \ erg}$

Nuclear chemistry

Ex.4 Nuclear reaction providing energy to sun and stars is as follows:

$$^2_1H + ^2_1H \rightarrow {}^4_2He$$

if ^2H= 2.01355 a.m.u. and ^4He = 4.00260 a.m.u., then how much energy is released when ^2H nucleus react to form ^4He nucleus ? If one mole of ^2H react and gives ^4He, then how much energy is released?

Solution:

Change in mass = Mass of reactant - Mass of product

Δm = 2(Mass of ^2_1H) - (Mass of ^4_2He)

= 2 x (2.01355) - (4.00260)

= 4.0271 - 4.00260

Am = 0.0245 a.m.u.

Released energy,

$\Delta E = \Delta m\ c^2$ where, c = 3.0 x 10^{10}

ΔE = 0.0245 x (3.0 x 10^{10})^2

ΔE = 0.2205 x 10^{20} a.m.u.

Now, 1 a.m.u. = 1.6 x 10^{-24} gm

ΔE = 0.2205 x 10^{20} x 1.6 x 10^{-24} erg.

= 0.3528 x 10^{-4} Erg.

= 3.528 x 10^{-5} Erg.

Now, 1 Joule = 10^{-7} Erg.

ΔE = 0.3528 x 10^{-5} x 10^{-7}

= 3.528 x 10^{-12} Joule

(c) if 1 mole of 2H reacts and lives ^{4}He which released energy,

$E = (\Delta E$ x N$)/2$ (N = Avogadro number = 6.02 x 10^{23})

= (3.528 x 10^{-12} x 6.02 x 10^{23} Joule

$E = 1.062$ x 10^{12} Joule

Ex. 5. Energetic proton strikes on $^{7}_{3}$Li and produce $^{4}_{2}$He particles. Calculate releasing energy during reaction in MeV.

(Li=7.01823 amu, H =1.00815 amu, He= 4.00387 amu)

Solution:

(a) $^{7}_{3}$Li $+^{1}_{1}$H $\rightarrow$ 2 $^{4}_{2}$He

Reduction (or decrease) in mass

Δm=Mass of reactant- mass of product

= ($^{7}_{3}$Li $+^{1}_{1}$H) -2($^{4}_{2}$He)

25

Nuclear chemistry

= (7.01823 + 1.00815) - 2 (4.00382)

= 8.02638- 8.00774 amu

Δm = 0.01864 a.m.u.

Decrease in mass= 0.01864 a.m.u.

(b) Energy E (in MeV),

E (MeV) =Δm(amu) x 931

= 0.01864 x 931

E = 17.3538 MeV.

Ex. 6. The physical atomic mass of two isotopes of magnesium Mg^{24} and Mg^{25} are accordingly (or orderly) 23.9924 a.m.u. and 23.9938 a.m.u. Calculate bonding energy of each.

Solution:

(1) Calculation of bonding energy of Mg^{24}:

Mg^{24} → $_{12}$Mg^{24}

It means, p = 12 and n = 12 z= 12 A = 24

Nov change in mass,

Δm= [Z$_{mp}$ + (A - Z)m$_n$] - m

= [12 x 1.0078 + (24 - 12) x 1.00871 - 23.9924

Nuclear chemistry

= [12 x 1.0078 + 12 x 1.0087] - 23.9924

= [12.0936 + 12.1044] - 23.9924

= 24.198 - 23.9924

= 0.2056 a.m.u.

So total bonding energy,

Δm= 0.2056 a.m.u. x **931 MeV/amu**

    = **191.41 MeV**

Mg^{25}: Calculation of bonding energy of Mg^{25}

In, Mg^{25} → $_{12}Mg^{25}$

    z= 12 and A = 25

Now change in mass,

Δm= [Z_{mp} + (A - Z)m_n] - m

    = [12x 1.0078 + (25 - 12) 1.0087] - 24.9938

    = [12.0936 + 13.1131] - 24.9938

    = 25.2067 - 24.9938

    = 0.2129 a.m.u.

Total bonding energy

Δm= 0.2129a.m.u. x 931 Mev/amu

    = **198.21 MeV**

Nuclear chemistry

Bibliography

1. Essentials of Nuclear Chemistry: Arnikar, 4th Edition (2012 reprint), New Age International.

2. Physical Chemistry: Atkins, 9th Edition. Oxford University Press.

3. Advanced Physical chemistry: Gurtu and Gurtu, 11th Edition , Pragati Prakashan.

4. Physical chemistry: Levine, 6th Edition, McGraw-Hill education, India.

5. Physical Chemistry: G. M. Barrow, 5th Edition, McGraw-Hill education, India.

6. Advanced Physical Chemistry: Gurdeep Raj, 35th Edition (2009), Goel / Krshina Publishing House.

Nuclear chemistry

Questions

- ❖ Definition: Isotops, Isomers, Isotones, Isobar.

- ❖ Explain in details about Bain-bridge velocity focusing mass spectrograph.

- ❖ Give in details about Nier's double focusing mass spectrograph.

- ❖ Write a note on Applications of Radioactive Isotopes.